Rain

Longman · York Press

Rain brings life to the earth.

It waters the ground
so that plants can grow.

Rain fills the lakes and rivers
where animals drink.

4

It fills up reservoirs and wells,
giving us water to use.

People, animals and plants all need water
so that they can live and grow.

We know that
rain comes from clouds in the sky.
But how is rain made?

Look at this saucepan filled with water.
When the water becomes very hot, it bubbles.
Steam comes out of the water
and rises into the air.

The sun heats the earth
just as the flame heats the saucepan.

The sun warms the water on the earth.
It rises into the air
like the steam from the saucepan.

10

As it rises into the air, it grows cold.
Then it changes back into tiny drops of water.
These are so small that they can float in the air
and form a cloud.

Later, the tiny drops join together
to make bigger drops.
These are too heavy to float in the air
and they fall to the earth as rain.

Sometimes the rain falls very heavily.

Some countries have too much rain
so they often have floods.

Because there is so much rain here,
forests like this are called rain-forests
or jungles.

Other countries have too little rain,
so they have deserts.

Camels can live happily in the desert because they can last for many days without water.

Even in the desert it rains from time to time.
Suddenly seeds lying in the sand begin to grow.
For a few days the desert is covered with
wild plants and flowers.

Where there is not enough rain,
water can be taken along canals or pipes
to help plants to grow.

Rain turns to snow if it is very cold.

Snow can be great fun.

It is so cold on top of some mountains
that the snow never goes away.

22

Rain can freeze and turn into ice.
Lakes and ponds freeze over
when it is very cold.

These people are skating on ice.

Look, it is raining in one part of the sky,
and the sun is shining in another.
If we turn our backs to the sun
we can see a rainbow.

There are many different kinds of clouds.

Sailors and farmers look at the clouds
to see how the weather will be.

Before it rains
the earth often looks dry and dusty.

28

After the rain has fallen
everything looks fresh and bright.

Rain makes the world beautiful.

Did you know that . . .

 The greatest amount of rainfall in a month
was recorded in India: 9,299 millimetres.
In the wettest months in Egypt there are only
about 5 millimetres of rain.

 In parts of Hawaii in the Pacific Ocean
it has been known to rain for 350 days in a year.

 There are some people called rainmakers
who are supposed to be able to cause rain.
Rain can also be made to fall in dry seasons
by spraying clouds with chemicals from an aeroplane.
This is called 'seeding the clouds'.

 Most rainbows only last for a few minutes.
But one rainbow in Wales lasted for over 3 hours.

 When the sky is very cold, rain can freeze
and fall as lumps of ice. This is called hail.
Hail stones can be big enough
to break a window or kill a chicken.

 No two snowflakes are ever the same.

Index

	page		page
air	8, 11, 12	pond	23
animal	4, 6	rainbow	25, 31
camel	17	rainfall	31
canal	19	rain-forest	15
chemicals	31	rainmaker	31
cloud	7, 11, 26, 27, 31	reservoir	5
country	14	river	4
desert	16, 17, 18	sailor	27
drop	11, 12	saucepan	8, 9, 10
earth	2, 9, 10, 12, 28	season	31
farmer	27	seed	18
flood	14	skating	24
flower	18	sky	7, 25
hail	31	snow	20, 21, 22
hailstone	31	snowflake	31
ice	23	steam	8, 10
jungle	15	sun	9, 10, 25
lake	4, 23	water	5, 6, 9, 11, 17, 19
pipe	19	weather	27
plant	3, 6, 18, 19	well	5

LONGMAN GROUP LIMITED, Burnt Mill, Harlow, Essex
YORK PRESS, Immeuble Esseily, Place Riad Solh, Beirut

Illustrations by Pat Robson

© Librairie du Liban 1985

First published 1985   ISBN 0 582 24500 1
Printed in Spain